WITHDRAWN

NAFC PUBLIC LIBRARY

Y513.21 HUNT
Hunt, Dawson J.
Farmers market measurements
M   394042   10/24/12
New Albany-Floyd Co. Library

Farmers Market Measurements

Dawson J. Hunt

CAPSTONE PRESS
a capstone imprint

First hardcover edition published in 2011 by
Capstone Press
151 Good Counsel Drive, P.O. Box 669, Mankato, MN 56002
www.capstonepub.com

Published in cooperation with Teacher Created Materials. Teacher Created Materials is a copyright owner of the content contained in this title.

 This book was manufactured with paper containing at least 10 percent post-consumer waste.

Editorial Credits

Dona Herweck Rice, editor-in-chief; Lee Aucoin, creative director; Sara Johnson, senior editor; Jamey Acosta, associate editor; Neri Garcia and Gene Bentdahl, designers; Stephanie Reid, photo editor; Rachelle Cracchiolo, M.A. Ed., publisher; Eric Manske, production specialist

Library of Congress Cataloging-in-Publication Data
Hunt, Dawson J.
 Farmers market measurements / by Dawson J. Hunt.
    p. cm.—(Real world math)
 Includes index.
 ISBN 978-1-4296-6846-0 (library binding)
 1. Arithmetic—Miscellanea—Juvenile literature. 2. Farmers' markets—Miscellanea—Juvenile literature. 3. Farms—Miscellanea—Juvenile literature. I. Title.
 QA115.H86 2011
 513.2′1—dc22   2011001575

Image Credits

Alamy/Emilio Ereza, 26; Richard Levine, 23 (left)
BigStockPhoto/Stefen Zwijsen, 27
Shutterstock/Ales Studeny, 9 (bottom left); Cathleen A Clapper, 13 (top); Christopher P. Grant, 19 (middle right); Colin Stitt, 19 (right); darkgreenwolf, 23 (right); Elena Elisseeva, 10; Eva Madrazo, 24 (top); Feng Yu, 13 (bottom); Heather Barr, 22 (back); HomeStudio, 9 (bottom right); Ibo, 16 (top); Jacob Kearns, cover, 1; James R. Martin, 21 (top, smaller pumpkins); Jim Hughes, 21 (bottom, scale); John Kroetch, 4; Joseph E. Ligori, 15; Laura Stone, 5; lfstewart, 12 (right); Maria Dryfhout, 8, 9 (top); Mario Savoia, 11 (top); Melanie DeFazio, 7; Michael C. Gray, 21 (top, large pumpkin); Morgan Lane Photography, 22 (front); Nadejda Ivanova, 19 (left); Nikola Bilic, 17; Paul Prescott, 14; photazz, 6 (all); Robyn Mackenzie, 25; Sebastian Kaulitzki, 21 (bottom, pumpkin); sf2301420max, 12 (left); Solov'ev Artem, 16 (bottom); Steven Pepple, 19 (middle left); Susan Fox, 19 (top); Thomas Barrat, 20; Tomo Jesenicnik, 18; Veronika Bakos, 11 (bottom); Viktor1, 24 (bottom)
Tim Bradley, 29 (all)

Printed in the United States of America in Stevens Point, Wisconsin.
032011   006111WZF11

Table of Contents

Spring 4

Summer 12

Fall 18

Winter 24

Problem-Solving Activity 28

Glossary 30

Index 31

Answer Key 32

Spring

It is springtime on our farm.
Each **season** is different on a farm.
But every Saturday is the same.

That is when we go to the farmers market. We all help sell what we have grown. It is the best day of the week.

In the spring we sell things that we grew in the winter. I help get the crops ready after school.

I like to put asparagus **spears** in bundles. I tie up 24 spears in each.

Let's Explore Math

1 dozen = 12 items

How many spears are in 2 dozen?

People like to buy our fresh peas. We sell snow peas and peas in pods. Buyers fill up bags with their favorite kinds of peas. Then we weigh the peas.

If the buyers choose peas in pods, they have some work to do. They have to get the peas out of the pods.

LET'S EXPLORE MATH

 =

1 pound of peas in pods = 1 cup of shelled peas

One family wants 4 cups of shelled peas. How many pounds of peas in pods should they buy?

My mom grows herbs to sell. We sell them in small bunches. My dad likes basil best. My sister likes mint.

I like dill best. Guess what my cat likes best? Catnip!

Mint can spread quickly. It sends out new parts just under the soil. These parts are called **runners**. Runners sprout new plants.

Summer

I help more on the farm in the summer. We grow lots of strawberries in a big field. My grandmother makes strawberry shortcake to sell. I add the whipped cream.

People love this sweet treat. We always sell out. I am glad we have more at home!

LET'S EXPLORE MATH

Strawberries are usually sold in a pint basket. This is called a **dry measure**.

There are 2 pints in 1 quart. How many pints are in 4 quarts?

Zucchini squash grows very fast. We pick ours at about 8 inches long. Once we missed one. It grew to be 2 feet long. It weighed 8 pounds!

Big zucchinis get very tough. But they are still good when made into zucchini bread.

By late summer we have a lot of corn to sell. We sell 10 ears for $1. When all of the corn is sold, I get to take a break.

A corn plant starts out as a seed. By July it may be 48 inches tall. That is nearly 122 centimeters. By the end of summer it may be twice as tall!

48 in.
122 cm

A band plays today. I get some fresh juice at a booth and sit down to listen.

Fall

We pick apples from our trees in the fall. We fill up baskets to bring to the farmers market. I cut up the apples so people can try a slice. Our apples are crisp and sweet.

Grandma makes apple pies too. They sell out fast!

LET'S EXPLORE MATH

Various tools are used for measuring. You are going to make an apple pie. Look at the measuring tools. Then answer the questions.

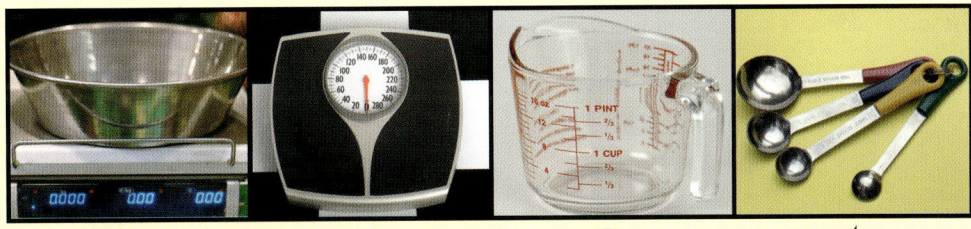

1.   2.   3.   4.

a. You need to buy some apples. Which tool would you use to measure them?

b. Which tool would you use to measure spices?

c. Which tool would you use to measure sugar?

The kids love our stand in the fall. We bring in lots of pumpkins. I like to line them up in various ways.

↑ The **height** of an object tells you how tall it is.

◯ The **circumference** is the distance around an item.

I may line them up from the shortest to the tallest. I may line them up from the thinnest to the thickest.

LET'S EXPLORE MATH

Pumpkins are sold by the pound. How much does this pumpkin weigh?

We sell **gourds** and Indian corn too. A gourd is a kind of squash. The corn is dried. People buy both items for their tables.

It is getting cooler. People have to dress warmly when they come to the farmers market in the fall.

LET'S EXPLORE MATH

The weather gets cooler in the fall. What temperature is shown on the thermometer?

Winter

We still have things to sell when it gets cold. Grandmother made strawberry jam in June. Dad filled jars with honey all summer. Mom and I made apple butter in the fall.

Bees create honey for their own food. They need it during cold weather.

We put samples of each on small biscuits. By noon we have sold it all.

Some days the weather is bad. The farmers market is closed on these days. But we are still busy. We may go to the **nursery** to buy new plants or tools.

Sometimes we just stay home and enjoy the fruits of our **labor**! Guess what food I like best?

Problem-Solving Activity

Selling Pumpkins

The Marks kids have 6 pumpkins. They are going to sell them to the Pumpkin Eater Cannery. The Pumpkin Eater Cannery buys pumpkins according to how tall they are. That is the height. This chart shows how much they pay for the pumpkins.

short pumpkins	$1
medium pumpkins	$2
tall pumpkins	$3

The Welch kids have 6 pumpkins too. They are going to sell them to the Peter Peter Cannery. The Peter Peter Cannery buys pumpkins according to how big around they are. That is their circumference. This chart shows how much they pay for the pumpkins.

small pumpkins	$1
medium pumpkins	$2
large pumpkins	$3

Solve It!

a. How much will the Marks kids earn?

b. How much will the Welch kids earn?

c. How else could the pumpkins be measured?

Use the steps below to help you solve the problems.

Step 1: Look at how tall the pumpkins are. Sort them into short, medium, and tall. Look at the chart. Add up what the Marks kids will earn.

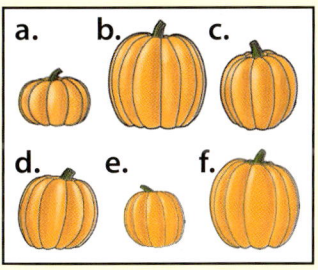

Step 2: Look at how big around the pumpkins are. Sort them into small, medium, and large. Look at the chart. Add up what the Welch kids will earn.

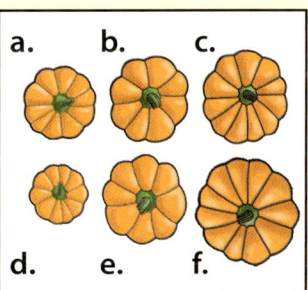

Glossary

bundles—groups of things tied together

circumference—the distance around an item

dry measure—a way to compare amounts of dry food items using containers

gourds—fruits with hard rinds that grow on vines

height—how tall something is

labor—work

nursery—a place where plants and trees are sold

runners—parts of a plant that grow under the soil and may grow into new plants

season—one of four time periods of the year

spears—growths of plant, usually young sprouts

Index

bunches, 10

circumference, 20

crops, 6

dozen, 7

farm, 4, 12

farmers market, 5, 18, 23, 26

height, 20

pound, 9, 14, 21

season, 4

weight, 8, 14, 21

ANSWER KEY

Let's Explore Math

Page 7:
24 spears

Page 9:
4 pounds

Page 13:
8 pints

Page 19:
a. 1
b. 4
c. 3

Page 21:
5 pounds

Page 23:
62°F (17°C)

Pages 28–29:
Problem-Solving Activity
a. $12
b. $12
c. The pumpkins could be measured by weight.